S0-AWD-794

"Ed Rosenthal wants cannabis fanciers to stop seeing their dealers. Not that he's a narc or anything—to the contrary. He just has a proposition that he thinks will help decommercialize the grass market.

"Rosenthal is offering the first grow-it-yourself grass environment. His six foot high constructions are supposed to be producing enough marijuana to keep a stash perpetually self-sufficient after about four months."

Rolling Stone 6/22/72

"Since Ed had been raving about the extraordinary quality of his home grown product, I called together an impartial and fairly knowledgeable panel to test his claim. Participants reacted with extreme enthusiasm, a majority insisting it was the best grass they had ever smoked. One in particular waxed ecstatic about the marvelous and amazing hallucinogenic qualities of the huge-leafed organic grass, but I can't vouch for her; she may have been stoned at the time."

New York *Village Voice* 6/24/71

The
Indoor Outdoor
Highest Quality

MARIJUANA
GROWER'S GUIDE

The
Indoor Outdoor
Highest Quality

MARIJUANA
GROWER'S GUIDE

by

Mel Frank and Ed Rosenthal

illustrations

Larry Todd

And/Or Press
San Francisco

This book is being published in California in 1974. At this particular point in time, possession and cultivation of marijuana is illegal in the state of California as well as many other parts of the world. The publishers cannot advocate breaking the law and suggest strongly that the cultivator check the manmade laws which govern the part of the planet in which he intends to carry out the operations described herein.

Level Press:

> first printing — June, 1974
>
> second printing — September, 1974
>
> third printing — February, 1975

And/Or Press:

> first printing — June, 1975
>
> second printing — October, 1975
>
> third printing — January, 1976
>
> fourth printing — April, 1976
>
> fifth printing — August, 1976

© *Copyright 1974*

Mel Frank and Ed Rosenthal.

AND/OR PRESS
3431 RINCON ANNEX
SAN FRANCISCO, CA
94119

Cover photo: BJM

Cover layout: C. Schnabel

Dedicated to
Sister Marie Louise Tibeau

FORWARD

Over the past ten years there have been revolutionary changes in the values of young people. The empty materialism of the fifties and liberal idealism of the sixties have been washed away by a pragmatic re-evaluation of lifestyle and political structure.

To a great extent this is the result of the widespread use of the psychoactive herbs and drugs which burst upon the scene in 1967. These substances seem to break down the ego and defense mechanisms and allow individuals to re-evaluate the sets and set perceptions based upon behavior patterns no longer revelant.

Marijuana, the most popular of the psycho-active herbs, has helped millions of people to a broader understanding of themselves. It is for precisely this reason that governments all over the world view it as a dangerous drug. How can they control their people if they see through the hypocrisy and self-serving purpose of the leaders' actions?

The use of marijuana has become so widespread that the government's repressive efforts, such as Operation Intercept, have resulted in almost total failure. and contempt for the inept efforts by the rulers of a system which hears its death-knell but does not understand the sounds.

However, the economic system which makes marijuana seem like just another commodity, rather than the sacrament that it should be, must be replaced. This book will help you make marijuana free

FREE GRASS   FREE YOURSELF
FREE THE WORLD

TABLE of Contents

Outdoor Cultivation

Clearlight Growing Systems

Light Charts

INTRODUCTION

Many of you at some time may have tried to grow marijuana. Very likely you buried some seeds in a flower pot, watered it and put it on a window sill. When the plants came up, you watered them faithfully every day. The plants, perhaps, grew for a few months to about 18 inches, at which point you cut them down and harvested four disappointing joints. It need not have been that way.

The purpose of this book is to show you how to grow high quality grass easily and cheaply no matter where you live. Using fluorescent lights, it is easy to grow very potent seven-foot, plants. Outdoors you will be able to harvest a large, potent crop with little effort. The experience would be well worth it even if you didn't harvest a joint. The process of learning to nurture and respond to your plants can be a very humanizing one. If you are already a plant grower, you will understand. If not, read through this book at a sitting, imagining the various stages of your plants' growth and the various decisions you, as grower, would be making to help your plants grow to full and healthy maturity.

An indoor garden, costing less than $50, will produce about 16 ounces of grass every six months. The gardens are simple to build. Basically all you need to do is hang a flourescent light that can be raised and lowered over some pots with a good soil mixture. All the materials you need are available at nurseries, garden shops, and hardware or lighting stores.

Homegrown grass is bright green and tastes pure and clean. Once you have experienced the pleasure of working with nature and enjoying clean, clear highs, we doubt that you'll ever buy commercial grass again.

Marijuana is an especially rewarding plant to cultivate because it is one of the fastest growing and most responsive plants. The hemp plant (cannabis sativa) is highly adaptive and grown around the world and, under optimum conditions, may attain a height of twenty feet. Such giants grow in tropical and semi-tropical zones where they flourish in the strong sunlight. There are male and female plants as well as hermaphrodites (male and female flowers on one plant.)

13

Indoor
Cultivation

THE INDOOR GARDEN

Under artificial light, marijuana grows very fast, about 3 to 6 feet in three months. You must be easily able to adjust the height of the lights as the plants grow. Hang the fixtures by a rope or chain from the walls or ceiling, or from the top of a frame **at least** 6 feet high constructed of 2" x 2"s. Try to obtain an industrial type fixture with a built-in reflector, so that no light is lost. If your fixtures are not equipped with reflectors, mount them on a sheet of white painted plywood or make reflectors with white posterboard.

Ten watts per square foot of growing area is adequate for healthy growth but for a fast growing, lush crop, use at least 20 watts per square foot. The dimensions of the garden should correspond to the light system, so if the garden is 1 by 4 feet, use two 4 foot tubes (80 watts). If the garden is 8 x 2 feet, use four 8 foot tubes (320 watts). If you have the space, use an 8 foot tube instead of two 4 foot tubes. One 8 foot tube emits more light than two 4 foot tubes. Marijuana can absorb up to 60 watts per square foot. Increasing the amount of light will increase the growth rate.

The garden should be surrounded by reflective surfaces to contain all of the light. This will increase the efficiency of the lighting significantly and the light will be nearly uniform throughout the garden until the fixtures are more than 2 feet high. A flat white paint (super white or decorator white) is a better reflecting surface than aluminum foil, mylar or glossy white paint. Flat white has about the same reflecting capacity as aluminum foil but reflects light more uniformly. Paint walls flat white and hang posterboard or white plastic curtains on any open sides from the top of the fixture or frame. Marijuana grows well in a dry climate but heated or air-conditioned homes are often too dry. Enclosing the garden with reflectors will insure a healthy humidity by containing moisture evaporating from the soil and transpired by the plants.

Don't rely on training pets to stay out of the garden. They may be attracted by the plants and chew the leaves and can break the stems. Soil is more natural to their instincts than the sidewalk or kitty litter tray. One moment of weakness can destroy months of work. If the garden is accessible to pets, surround it with chicken wire or heavy plastic. Hardware stores sell plastic on rolls and inexpensive plastic dropcloths. Cover the floor with plastic too. It will protect your floor and your neighbor's ceiling from possible water damage.

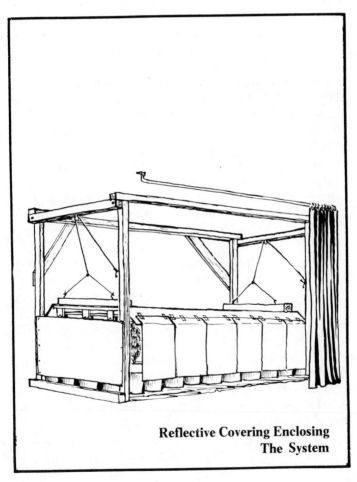

**Reflective Covering Enclosing
The System**

ARTIFICIAL LIGHT

The most effective and efficient artificial light for plant growth is fluorescent light. The white light you see emmitted by a fluorescent bulb consists of all the colors of the light spectrum. The designation—Daylight, Warm white, Gro-Lux, Optima etc.—corresponds to a particular combination of light generated in each of the color bands. Plants respond primarily to red and blue light and for healthy growth a combination of these colors must be provided. Blue light stimulates leaf growth, produces short, stocky stems and encourages robust development. Red light is used for stem and root growth and to promote flowering.

The best flourescent tubes to use are those specifically manufactured for plant growth. Some of these are the Standard Gro-Lux, Wide spectrum Gro-Lux, and Plant-Gro. Vita-Lite, Optima and Natur-Escent were originally intended to human vision but are now widely used for plant growth and will work at least as well as the gro-tubes. Gro-Lux and Plant-Gro tubes look purplish and emit mostly red and blue waves. Wide spectrum Gro-Lux emits more ultra-violet but less blue and red waves than the Standard Gro-Lux. Vita-Lite, Optima and Natur-Escent look bright white. Their spectrum closely resembles that of natural light.

Sizes suitable for growing marijuana are 4, 6 and 8 feet long. Regular wattages for all of these tubes are about 10 watts for each foot of their length (80 watts for an 8 foot tube). They also come in higher wattage sizes. These are power twists, high output (HO) and very high output (VHO) tubes, the largest being a 215 watt, 8 foot, VHO bulb. These high output tubes are not always available and are quite a bit more expensive as are their fixtures. These tubes should be used if you are building a large garden. (See **Larger Systems**). Regular fluorescent tubes can be used if you can't get the Gro-Tubes. They don't work quite as well but they will grow a healthy crop and are sometimes cheaper. Daylight, or Cool White tubes can be used in combination with either Natural White or the standard incandescent (screw-in) bulb. Incandescents and Natural White both provide a strong red spectrum while the others tend more towards blue. Use them in a one to one ratio, evenly distributing the red and blue sources. Compared to the fluorescent tube, the incandescent bulb is about one third as efficient, has a much shorter life, and can create problems because of concentrated heat, which can burn the plants. Combining fluorescent with incandescent light can work well but it is better to use the Natural White fluorescents as the red spectrum source.

Cost for electricity will be about $3 or $4 per month for a one foot by eight foot (160 watts) system.

For more information on this subject see **Lighting for Plant Growth** by Bickford and Dunn, Kent State University Press.

See Light Charts pages 92, 93, 94.

Plant responses to different color bands		
Nanometers	**Color**	**Effects**
380 and shorter	Ultra-Violet	detrimental to life
380 - 440	Violet	some formative effects
440 - 500	Blue	secondary peak in photosynthesis, strong formative effects.
500 - 620	Yellow, Green, Orange	minimum formative effects
620 - 700	Red	primary peak in photosynthesis, maximum formative effects
700 - 1000	Far-Red	elongation
1000 and longer	Infra-Red	Heat only

POTS AND CONTAINERS

Plants can be started in flower pots, milk crates, institutional size cans (often available from restaurants), plastic jugs, bushel baskets or practically any container that can withstand repeated waterings and is at least four inches wide. Holes must be punched in the bottom of all containers to assure adequate drainage.

Plastic pots can be purchased from your local nursery. Many sell used pots for a few cents each. Wholesalers sell plastic pots by the carton for a reduced cost. A single large box or trough has the advantage of minimal restriction of roots and fewer waterings but require much more soil and makes rotating or moving the plants impractical.

In its natural state marijuana sends down a tap root up to half the length of the plant. Trying to simulate natural conditions would be impractical in terms of weight, space, cost, and labor.

The purpose of the growing medium is to provide adequate water and nutrients and to anchor the roots. With strict attention to proper watering and fertilizing techniques a six foot plant can be grown in a four inch pot. The plants will grow better in successively larger pots—six to ten inch pots are good median sizes to work with.

Use as many pots as you can fit under the light system. The pots can always be thinned if the plants become too crowded. Choose pots that are at least as wide at the top as at the bottom so that the soil can breathe and dry out more easily.

Wash all cans, containers or used pots thoroughly to remove any contaminants. Boil used clay pots for ten minutes to sterilize them.

SOIL PREPARATION

Marijuana grows best in a well drained sandy soil or loam which is high in nitrogen and potash, and at least medium in phosphorous, and which contains little or no clay. The pH should be between slightly acidic (6.5) to slightly alkaline (7.5). If the pH is either too low or too high it will interfere with nutrient uptake. The pH is measured on a scale of 0 to 14 with 7.0 assigned as neutral, and is a measure of relative concentrations of hydronium ions (H_2O) to hydroxide ions (OH_2). A simple and inexpensive test kit can be purchased at most garden shops to test pH and nutrient content. Many state agricultural schools or services will test pH and nutrients content for a nominal fee.

Soil pH is raised to an acceptable level by adding hydrated lime, limestone or marl. There is no set formula we can give you for raising the pH. At low pH it takes less lime to raise the level 1 point than it does when the pH is nearer to neutral. Sandy soils require less lime than clay soils to raise the pH. In general, if the soil tests acid, add 2 cups of hydrated lime for each 50 lb. bag of soil. This works out to about 1 tablespoon for every 1½ lbs. soil. Wet the soil thoroughly. Retest the soil in about 2 weeks and repeat the application until the pH is in an acceptable range. Soil that is too alkaline is treated the same way using aluminum sulfate at a rate of ½ cup per 50 lbs. of soil.

If you are digging up your soil, sift it to remove stones and root clods. Bake the soil in a 220 degree oven twenty minutes in 1 inch layers or in a pressure cooker at 15 lbs. pressure to destroy any weed seeds and insect eggs.

It is simpler and quite inexpensive to buy commercially prepared soils. These are usually sterilized and have a good balance of nutrients. Ask for soil with a neutral pH. Some nurserymen will sell you anything so it is advisable to test the pH level.

Soil consistency is important for healthy root development, drainage, and uniform water dispersion. The medium should not cake when dry and should remain spongy or loose when wet. To test consistency, lightly moisten the soil and compact it in your fist. The ball should crumble easily when touched. Soil consistency can be adjusted by adding perlite, vermiculite, sand or kitty litter.

Perlite and vermiculite are inexpensive commercial products which are much lighter than sand and are sterile. Vermiculite absorbs and holds water and air in its fiber. Perlite traps moisture and air on its irregular surface much like sand. Sphagnum or peat moss is often used to adjust soils, but should not be used for marijuana as it tends to make the mix to acidic.

Soils found deficient in nutrient content can be enriched by adding humus (decayed organic matter) or fertilizers such as sterilized cow manure and chicken dung or prepared fertilizers such as Electra-Mix, rose food, etc. Humus is acidic and can alter the pH.

Soilless mixtures are inexpensive and easy to prepare. They work well, are neutral in pH and light in weight, but have no nutrient content. They must be carefully fertilized and are not recommended for an inexperienced grower. Two tested formulas are:

1) 1 part perlite or sand to one part vermiculite and 1 tbsp. of lime per quart of mixture.

2) 1 part perlite or sand to 1 part Jiffy mix and 1 tbsp. of lime per quart mixture.

You can mix 3 parts of the soilless mixture to 1 part cow manure or rely solely on soluble fertilizers when watering.

Some of you who are not familiar with gardening may be a little over-whelmed by all this talk of pH, nitrogen, etc...so here is a simple uncomplicated formula for those of you with no experience with plants.

Buy commercial soil. Avoid brands that have "peat" in their names. It is very unlikely that a commercial soil will be too alkaline for healthy growth, but it may be too acidic. The simplest way to assure yourself that the soil is not too acidic is with blue litmus paper (available at any drugstore for about 25 cents). To test, take a tab of litmus and cover about half-way with wet soil. Wait a few minutes and remove. If the paper turns pink, the soil is acidic and lime must be added. Mix 2 cups of hydrated lime (from hardware and nursery stores) to each 50 lbs. of soil. Don't add lime if the litmus paper remains blue. If you find yourself asking, "is this paper pink, purple or just wet?" then the soil is probably slightly acidic and within an acceptable range for healthy growth.

Mix 10 lbs. of natural, sterilized fertilizer (cow manure or chicken dung) to each 50 lbs. of soil. The fertilizer does not have to be added if you plan to use soluble fertilizer when watering. By volume, mix 3 parts soil with 1 part of an adjuster (perlite, sand or vermiculite).

After potting and watering, the mixture should be retested in about 2 weeks if it tested acidic to begin with. If it is still acidic, add hydrated lime by mixing 1 tbsp. of lime per quart of water, the first few times you water. Test your water supply by dipping litmus paper in plain water to determine whether it is influencing your tests.

To pot any of the mixtures, cover the drainage holes with a square of newspaper or window screen to prevent soil from running out. Place a layer of sand, perlite, vermiculite or kitty litter about one inch deep to insure good drainage. Fill the pots to within 3/4 inch from the top of the pot with the soil mixture. Water the pots until the soil is evenly moist and allow the pots to stand for a day or two so that bacteria necessary for nutrient uptake can begin to grow and the fertilizers can start to dissolve.

For more information on this subject see **Soils** by Donahue/Shickluna/Robertson, Prentice-Hall, Inc.

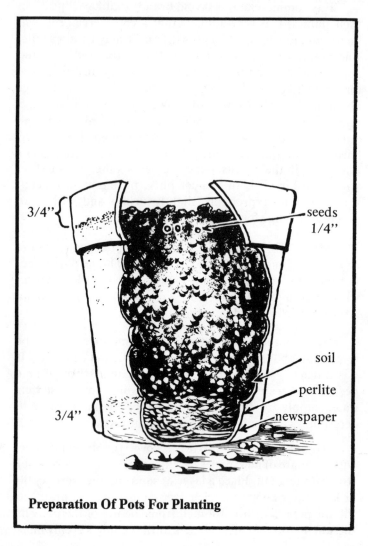

3/4"

seeds
1/4"

soil

perlite

newspaper

3/4"

Preparation Of Pots For Planting

24

SEEDS

The potency of marijuana is, in part, hereditary. Choose your seeds from the best grass available. Different strains grow at different rates. For uniformity of growth, take all seeds from the same batch of grass. Choose seeds for their size and color. The large plump ones with good color—black, brown, grey or mottled—have the best chance of germinating. Seeds that are old, badly bruised or immature (green or white) are probably not viable. Seeds are rarely viable after three years and should be stored in a cool, dark place in an airtight container. The vegetable crisper section of your refrigerator is an ideal place. You can get some indication of the viability by placing a seed between thumb and forefinger. If the seed does not crumble when pressed hard, it is probably viable.

Many books recommend that a germination box be built to start the seeds. This is an extra hassle that is not necessary. Transplanting the seedlings from one medium to another often subjects them to transplant shock which will delay growth. With the following procedure you'll have not problems.

Soak the seeds overnight in a glass of water or wet towels to give them a head start in water absorption. Adding about 2 teaspoons of bleach to each cup of water (a 5% solution) will prevent fungus from forming on the seeds. Poke five or six holes about ¼ to ½ inch deep and evenly spaced in each pot. Place one seed in each hole and cover lightly with soil. Carefully, so as to not disturb the seeds, moisten the soil and keep it moist until the seeds have sprouted. The seeds will sprout in three to fourteen days depending upon their variety and viability. If you have only a few seeds and want to give them the best chance possible, plant them pointed end up. The seedling will then expend the least amount of energy breaking through the soil. This is not critical and is unnecessary if you have plenty of seeds.

Place Seeds 1/4 - 1/2 Inch Deep

LIGHT SYSTEM
AND GERMINATION

Hang the lights as close as possible to the top of the pots during germination. Leave the lights on 24 hours a day. The close distance and long cycle will heat the soil and encourage germination.

Once the seeds have sprouted raise the light two to six inches above the top of the plant tops and maintain this distance for the duration of growth. The short distance between the light and plant will encourage the seedling to develop with a short stocky stem rather than a long fragile one reaching for the light. At some stages the plants will grow a couple of inches a day, so you may have to adjust the lights several times a week. Usually seeds will sprout 2-7 days after planting. Older seeds may take up to 3 weeks to sprout.

It is important for the normal development of the plants that they receive a regulated day/night cycle. We emphatically recommend that you use an automatic electric timer (about $10.00) so your plants will not suffer your irregular hours or weekend vacations. Once the seeds begin to sprout set the timer cycle for 18 hours of light a day, and leave it on this setting for the duration of your garden.*

It is best to set the timer so that your plants are not distrubed by any light during their night period. If they are subjected to even dim light too often during the night cycle the plants' growth pattern may be disrupted and they may develop abnormally. If you must use a light in the growing area, use a green bulb. Plants are not sensitive to the green spectrum.

*See Photoperiod section

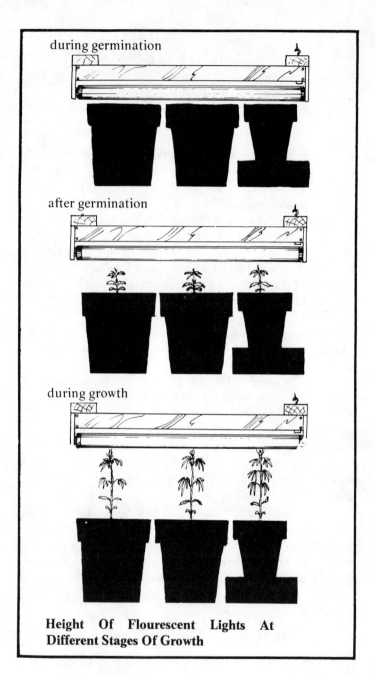

during germination

after germination

during growth

Height Of Flourescent Lights At Different Stages Of Growth

WATERING

Plants under artificial light have a long photoperiod and no cloudy days so they grow extremely fast; one and a half feet per month is not unusual. This means the plants will use a lot of water. Since the space around your plants is limited, you'll have to water often. This does not mean watering daily or keeping the pots saturated. Plants grown in a continuously wet soil are slower growing and probably less potent than normal and often develop stem rot. Allow the pots to go through a wet and dry cycle. This will aid in nutrient uptake, especially potassium, and aerate the soil. In general, when the soil one inch deep is dry to the touch, water enough so that the soil is saturated but not so much that water runs out of the drainage holes carrying away the soil's nutrients.

If you underwater, the plant will wilt. Plant cells are kept rigid by pressure of the cell contents (mostly water). With the water gone, they collapse. First the bottom leaves droop and this condition works its way up the plant until the top lops over. If a plant wilts, water immediately and it will recover within a few minutes. This happens so fast you can actually follow the movement of water as it goes up the plant.

There is no way we can tell you exactly how often to water. Light temperature, humidity, and size of plants and pots are only a few of many variables that determine water uptake.

Try to reach a median. Don't keep the pots constantly moist and don't wait until the plant wilts. Use some common sense. A 6-foot plant in a 4-inch pot will have to be saturated almost every day. Large containers (10" width or more) should not be watered to saturation especially when the plants are small. They will not dry out quickly. Clay pots are porous and "breathe". They require more water than plastic or metal pots.

Don't disturb the roots when you water. Water around the stems, not on them. Seedlings are likely to fall over if watered roughly. Use a hand sprinkler for seedlings.

Use tepid water; it soaks into the soil more easily and will not shock the roots. Try to water during the plants' morning hours. Water from the top of the pot. If you do want to water from the bottom with trays, place a layer of pebbles or gravel in the trays to insure drainage. Don't let the pots sit in water until the soil becomes supersaturated. This prevents oxygen uptake and the plants will grow poorly.

Water in some areas is acidic (sulphurous) or alkaline (limestone) and can change soil pH.

Tap water is some locales is highly chlorinated. The chlorine does not harm the plants but it can kill the micro-organisms in the soil that are necessay to break down nutrients to a form that the plants can use. Allowing chlorinated water to stand overnight will eliminate most of the chlorine gas and it can then be used safely.

Water Round The Stems, Not On Them

THINNING

Depending on the viability of the seeds you will have a germination rate of 0-100% and several plants should be growing in each pot. During the 2nd to 4th week of growth the plants will begin to crowd each other. Thin your garden so that one plant is left in each container.

The marijuana leaf consists of 3 - 11 lanceolate shaped blades. These appear usually in odd numbers and the number depends on genetic factors and growing conditions, principally the amount of light. The number of blades at the early stage is an indication of over all leafiness at maturity.

To thin your garden, remove any plants with yellow, white or distorted leaves. Also remove the less vigorous ones and those with the sparcest foilage. Leave the bushiest and those with the highest number of blades per leaf.

If the plants are close together, cut the unwanted plants at their base: the root system can remain in the pot. Otherwise make sure that you do not disturb the remaining plants' root systems when you pull the unwanted seedlings. The tops of these harvested plants will be your first taste of homegrown grass. They will produce a mild buzz. Potency of the crop will increase considerably as the plants grow older.

TRANSPLANTING

If there are any pots without plants, you should transplant a seedling into it when you are thinning.

First, moisten the soil in the pot from which you will remove the transplant and let it sit for a few minutes. Take a spade or a large spoon and inset it between the transplant

Do Not Disturb Root System When Transplanting

and the plant that will be left to grow. Try to leave at least one inch of space from spoon to stem. Lever the spoon towards the side of the pot so as to take up a good size wedge of the soil. Place the transplant in a prepared hole at the same depth that it was growing before. Replace the soil in both pots and moisten lightly again to bond the new soil with the original. If carefully done, a wedge of soil can be removed intact, so the root system will not be disturbed and the plant will survive with little or no transplant shock. Do not fertilize a transplant for two weeks.

To prevent drop-off and wilting from shock you may want to use Rootone or Transplantone. These powders, available at nurseries, contain root growth hormones and fungicides and are safe.

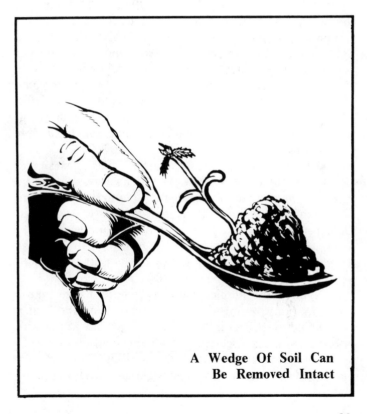

A Wedge Of Soil Can Be Removed Intact

SUPPORTS

Plants grown under artificial light will often need support, especially in the early stages of growth. Unlike sunlight on Earth, the intensity of artificial light diminishes the further the distance from the light source. The plants respond by trying to grow up to the light. Hanging the lights higher than the recommended 6 inches will further complicate this elongation. Too much red light will cause elongation too, so make sure that you include a strong blue light if you are using incandescent bulbs. The blue band will ease elongation somewhat but the heaviest foliage will still be on the top and the stem may not be able to support the weight.

Depending on plant size, Q-tips, plastic straws or standard plant stakes such as cane sticks or dowels can be used. Set them in the soil and tie the stem to it with a string or wire twists like those that come with plastic trash bags. Do not pull the string tightly around the stem; leave it very loose. Marijuana is a dicot and will grow in girth as well as length. Tying the string too tightly can cut off the flow of water and nutrients as the stem grows larger.

Probably the simplest method of support is to take a rigid piece of wire, form a C at one end, bend the C at a right angle to the stem, push the straight end of the wire into the ground and carefully place the stem inside the C. Wire pipe cleaners are ideal for seedlings. With larger plants, coat hangers can be straightened and the same method used.

34

Various Plant Supports

FERTILIZING

As the plants grow they take nutrients from the soil. These nutrients must be replaced if the plants are to stay healthy and strong. The main elements a plant uses are nitrogen (N), phosporous (P), and potassium (K). These are the three numbers listed on fertilizer packages: for example 5-10-5. Calcium, magnesium, sulphur, and iron are used in lesser amounts as are minute quantities of many other elements called trace elements or micro-nutrients. Each element affects different characteristics in the plant and all are necessary for healthy growth.

Nitrogen promotes rapid growth, lush foliage and stocky plants. During the first few months of growth, marijuana needs a lot of nitrogen. An abundance of nitrogen during the early stages will induce more female plants to develop. More males will develop if there is a lack of nitrogen during early growth stages.

Phosphorous promotes root growth and is necessary for healthy flower and seed development. When flowering, marijuana uses about twice as much phosphorous as it does during vegetative growth.

Potassium regulates nutrient utilization, increases vigor, strengthens stems and resistance to diseases and is essential for proper plant metabolism. The plant needs large amounts of potassium during all stages of growth.

Calcium aids in absorption of nutrients, neutralizes soil acids and toxic compounds produced by the plant.

Testing the soil periodically is the surest way of maintaining a healthy growing medium. Soil that tests high in nitrogen and potassium, medium in phosphorous will not have to be fertilized for a while. Soils found deficient in one element can be treated with a single component fertilizer.

When and how often to fertilize depends on the growing medium you started with, the size of pots and general growing conditions. Normally, small pots (4 to 6 inches) should be fertilized about three weeks after sprouting. Fish emulsion (5-2-2) is a good organic fertilizer. Dilute 1 teaspoon per gallon of water and use each time you water for the first two months and once every two weeks thereafter. Chemical fertilizers such as Rapid-Gro (23-19-71), Miracle-Gro (15-30-15) or Hyponex (many ratios offered) can be used in accordance with instructions listed for houseplants. Don't use fertilizers recommended for "acid loving plants," and never ad solid fertilizers like cow manure once the plants have started. They promote molds that can be harmful.

Large pots (10 to 18 inches) may not need to be fertilized at all if the soil was rich in nutrients to begin with.

Soilless mixtures must be treated with a trace element mixture. Mix one tablespoon per gallon of water the first time you water. Every six weeks water with one teaspoon per gallon. Increase the treatment if the plants show any trace element deficiencies. You can use any houseplant fertilizer such as Rapid-Gro, Miracle-Gro or Hyponex. These also contain trace elements. An ideal formulas ratio for producing the most desireable results at each of the plant's stages of life (rapid growth and profuse foliage in the beginning, good development during middle life, and high resin content during flowering) is as follows:

N - Nitrogen   P - Phosphorus   K - Potassium

	N	P	K
start in 2nd week	20	5	15
start 2 months before flowering	10	5	15
during flowering	5	12	10

It is not essential that you fertilize in these ratios, only that the plants receive enough of each element.

Use one tablespoon of micronutrient mix to each gallon of water during the first week and thereafter, use once every six weeks.

One week after sprouting, water with fertilizer in dilutions recommended on packages for large bushes, tomatoes, etc. Repeat this application in the third, fifth and eighth weeks. Thereafter fertilize in dilutions recommended for houseplants once every two weeks until flowering.

Some words of caution: many people in an effort to do the best for their plants actually do the worst. Overfertilizing will put excessive amounts of soluble substances into the soil. They interfere with normal nutritional processes and will cause poor growth and, in some cases, kill the plants. For example, too much nitrogen will nitrify the soil and change its osmotic properties. Instead of moisture being drawn into the plant it is drawn away and the plant dehydrates. In the limited area that your plant is occupying it is easy to overfertilize. If the plants look healthy and are growing well, don't be anxious to fertilize. It is better that they are underfed rather than overfed. Underfeeding can be recognized and corrected but overfertilization means you must start another crop and replace or leach the soil mix. It is better to use a more diluted solution more often than to give one large dose once a month.

Foliar feeding (spraying the leaves with fertilizer) is a good way to assure the plants their nutrients without building up soluble substances in the soil. After the first month, foliar feed the plants with fish emulsion or the chemical fertilizers. Some of the chemical fertilizers say, "not recommended for foliar feeding houseplants." Marijuana is not a houseplant. As long as the fertilizer can be used for foliar feeding, use it on your plants. Use a fine mist sprayer such as a clean Windex or Fantastic bottle. Dilute fish emulsion 1 teaspoon to a gallon and use each time you water. Spraying with fish emulsion is a little smelly but it may change the taste of the grass to a pleasant mint-like flavor. Dilute the chemical fertilizers according to package directions and spray weekly. The nutrients are absorbed through the leaf surface (both sides) and through "breathing holes" (stomata) in the leaves. Occasionally spray with plain water to redilute unabsorbed nutrients and to clean the plants.

If any plant has an unhealthy or discolored appearance, make sure the problem is not due to insects or disease before assuming a nutrient deficiency. Examine the plants carefully, especially the undersides of leaves, along the stem and in the soil.

Deficiency signs:

Nitrogen—Plant color is paler than normal. Yellowing of older leaves on the main stem followed by yellowing of younger leaves with slow or no growth. Yellowing of the bottom leaves will occur after the plant is more than 2½' tall since they are shielded by the upper leaves or are too far from the lights to carry on chlorosynthesis.

Phosphorous—Leaves are unnaturally dark green with slow growth. Poor flowering and root structure.

Potassium—Leaves are unnaturally dark green and curl under at the edges. Bronzing or yellowing starting on

39

the edges of older, main stem leaves, which then turn grey; followed by grey or bronze mottling of the whole leaf. Stems are often soft and weak. This is the most common deficiency in indoor gardens.

The following deficiencies are not common, especially if you are using fertilizers. If the plants are growing poorly, check the soil pH and drainage. If the water stays in a pool and takes more than a minute or so to be absorbed, then the soil is not draining properly. Leaves will brown at the tips, turn pale or yellow and severely curl.

Calcium—Growing tips wither and wilt. Buds may not develop.

Sulfur—Young leaves have veins of light green.

Magnesium—Older leaves are pale green or yellow; this soon spreads to the whole plant.

Iron—Young leaves are light green or yellow. Veins are darker green than surrounding tissue giving leaves a varicose vein appearance.

Boron—Young leaves are constricted and light green;

Zinc—Abnormally small leaves with yellow or wrinkled edges. Sometimes spotted. Sparce foliage, often having leaves only at the top of the plant.

Manganese—Bleached out spots on the leaves.

Chlorine—General yellowing of leaves that turn copper or orange. Roots are swollen at the ends.

Molybdenium—Young leaves are distorted. Sometimes there is a yellowing of leaves in the middle part of the plant.

Organic Way to Plant Protection, Emmaus Pa., Rodale Books, Inc., 1966.

FLOWERING

It is virtually impossible to recognize the gender of marijuana plants until they begin to flower. The male plant is usually the taller and matures in 3 to 5 months. Two weeks prior to flowering it will grow very fast (internodes elongate), then shoots will sprout with clusters of small, dangling white, greenish white, yellow or purplish flowers that hang from tiny side branches along the main stem, on branches and at the top of the main stem. When mature, the flowers open and a yellow anther protrudes and wind disperses the pollen.

The female plant, although shorter, is fuller with more complex branching and often twice as many leaves as the male. Her flower consists of a delicate, downy white stigma raised in a "V" sign which is attached at the base to an ovary that looks like a tiny green pod. If fertilized, one seed will develop in the ovary. When allowed to grow, the flowers develop into clusters or "cones" which are interspersed with small green leaves known as bracts. The female is the more desirable plant for marijuana cultivators since it produces many more leaves and is considerably more potent than the male.

Normally, male to female ratio in marijuana is about 1 to 1. Genetic and environmental conditions interact to determine gender. A strong light source, long photoperiod, abundant nitrogen in early growth and much spacing between plants stimulate female development. Poor growing conditions in general, such as weak light, low nutrient availability, short or erratic photoperiod and crowded conditions will produce more males.

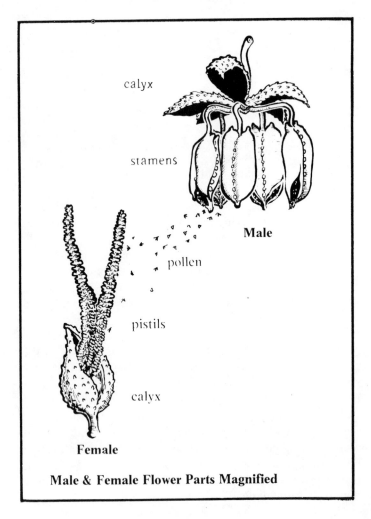

calyx

stamens

Male

pollen

pistils

calyx

Female

Male & Female Flower Parts Magnified

ROTATION AND EVEN GROWTH

The light intensity from artificial light drops dramatically as distance from the light source increases. If you don't keep the plants at about the same height, the shorter plants will receive less light and consequently will grow more slowly than the tall ones. This will compound the problem.

One way to deal with uneven growth is to line the plants up by height and hang the light system at an angle corresponding to the line of the plant tops.

If a few of your plants are markedly outgrowing the others, cut the growing tip back to the height of the average plant. You may find this emotionally difficult but it is important to the over-all health of your crop. Cutting the top will not hurt it but will force side branches to develop. Conversely, if a few plants are much shorter raise them by placing them on milk crates, cans or bricks, etc.

Young plants about two weeks old can be cut back. This forces branches to develop early and will quickly fill any available horizontal space. It is helpful with large pots where there is much space between young plants. Growing tips of branches can be cut back to force more branching. This produces a stout, bushy plant and provides an immediate supply of grass. Don't overdo it. Severe pruning can cause plants to develop into males.

The growing tip of the plant (apical meristem) contains an inhibitor that prevents the branches (lateral buds) from growing. The futher a lateral bud is from the growing tip, the less the effect of the inhibitor. This is why some species of plants form in a cone or Christmas tree shape. Under artificial light the bottom branches don't receive enough light to grow even though they are far enough from the inhibitor. Once the tip is removed, the next highest growing tip will be the source of the inhibitor.

Some growers just hate to cut the growing tip. It becomes the biggest and most potent "cone" at harvest time. To save the tip, control height and force branching, bend the top of the stem down in an arc and secure it with string or wire twists. This will neutralize the effects of the inhibitor somewhat and still maintain a strong growing tip. The string or wire twist should be removed after a couple of

Line The Plants Up By Height

days so the stem will not break itself by twisting up toward the light source.

The quality and quantity of light emitted by a fluorescent is not uniform along the length of the tube. There is more light at the center than at the ends. Female plants require more light than males. She is the more potent plant and should be given the best care. Once the plant's sex shows, move the males to the ends of the system leaving the stronger middle light for the females.

See Light Charts pages 89-94.

**Bend The Top Of The Stem Down And
Secure It With String Or Wire Twists**

PHOTOPERIOD

Many plant functions are regulated by the quantity and quality of light and the length of the photoperiod (daylength). Marijuana is a short day (long night) plant. The female produces flowers only when she senses the decrease of daylength. In the autumn the shortening day is her signal to flower and produce seeds for next year's crop before winter sets in. Flowering in the male does not depend on changes in the photoperiod. It flowers regardless of daylength in 3 to 5 months, depending upon the variety.

Although termed "short day," it is during the night period that the chemical reactions that control flowering occur if given a long enough and uninterrupted dark period. The dark period must be constant and at least 9 hours long for the chemical buildup to be completed. By changing the light period to less than 13 hours a day, the female responds by flowering profusely in about 2 or 3 weeks. Females grown with a daylength of 16 or more hours may flower but will do so sparcely and will not develop large flower clusters. The longer the photoperiod the more pronounced this effect.

Before flowering, the leaf growth will be very fast. Once flowering begins, the plant's energy goes to producing the flowers and the leaf growth slows. With this in mind, you can manipulate the photoperiod for either a continuously growing, vegetative state or for flowering and a harvest crop.

The continuous growth system emphasizes leaf growth and a continuous supply of grass. You can harvest the first grass, which will give you a buzz or better, in about two months and have a steady supply of potent grass after about four months. A one by four foot system will supply several joints a day. The grass is not quite as potent as with the harvest system but will be of excellent quality and favorably comparable to most commercial pot. The system is easy to care for and supplies a large amount of grass over a period of time.

The harvest method produces a crop every 4 to 9 months. The grass is very potent and is at least as good as the best commercial pot. Although you may gather a few leaves now and then, you'll have to wait until the crop is harvested for a large supply. The system should produce a minimum of one ounce of pot for each square foot of growing area.

See Light Charts pages 89-94.

CONTINUOUS GROWTH SYSTEM

Use Vita-Lite, Optima, Naturescent, Wide Spectrum Gro-Lux or combine Plant-Gro or Gro-Lux in one to one ratio with Daylight Tubes. The abundance of blue light will emphasize leaf growth and not flowering. Do not use incandescent bulbs. The photoperiod should be kept constant at eighteen hours of light a day for the duration of the garden.

After two months the plants will be stocky and the area filled with foliage. At this time bottom leaves begin to yellow because they are shielded from the light or are too far away from it to carry on photosynthesis. Pick any leaf as soon as it begins to yellow. Green leaves can also be picked sparingly along with some leaf buds.

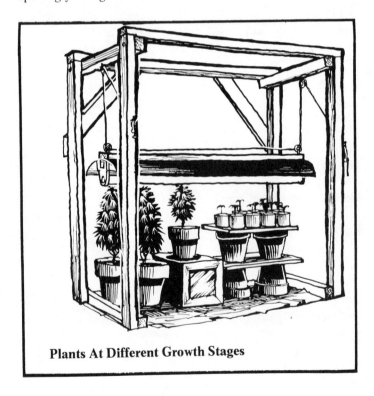

Plants At Different Growth Stages

Flowers may develop after four months on a few plants and can be picked just above the growing tip. New flowers will soon develop. Continue picking the flowers until the plant loses its vitality. Females usually will continue to grow for more than a year, but may lose their vitality after 8 or 9 months. When a plant's health begins to decline it should

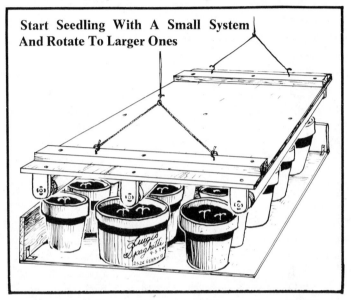

Start Seedling With A Small System And Rotate To Larger Ones

be uprooted and new plants started in its place. Seeds can be started or cuttings taken 3 inches below the growing tip of healthy plants. Use cuttings when you have an especially fast-growing or potent plant. Root the cutting directly in the soil using a transplant hormone such as Rootone or Transplantone. Expect a survival rate of 30 to 50 percent. Do not fertilize the cuttings for about two weeks. The light system will be quite high at this time so place the pots on milk crates or some sort of platform. In this way your garden will be kept in a continuous growing state with plants at different stages of growth giving you a constant supply of potent grass.

If you decide to start over completely or close the garden down, adjust the photoperiod accordingly and convert to a harvest crop.

HARVEST SYSTEM

Under natural conditions, the female plant adjusts its flowering to the length of the growing season. This is generally between 3 and 7 months, depending upon region and time of planting. Once the plants go to seed, they usually lose their vitality and soon die. Since you will be controlling the flowering mechanism, your females can be anywhere from 2 to 9 months old at harvest time. The potency of grass in general increases with age as long as the plant stays healthy. We have found that a happy medium in terms of potency and yield is to harvest about every 6 months.

Keep the photoperiod constant at 18 hours of light a day until 6 to 8 weeks before you plan to harvest. Then cut the day cycle down to about 13 hours of light. In about 2 weeks, the females will respond to the longer night and flowers will begin to appear. Allow the flowers to grow another 4 to 6 weeks so they can develop into large clusters which are by far the most potent part of the plant. Flowers can be harvested 2 or 3 times before uprooting the plant. Pick them just above their growing point where they meet the main leaves. New flowers will grow from this point giving you a higher yield of top quality grass.

Once the flowers have developed you might try a sunlamp for an hour or two a day at a distance of 3 feet to force resin to the flowering parts. The resin flow is the plant's protection against the intense heat and possibly the ultra-violet rays. The resin contains the cannabinols (THC) that make you high.

There is some discussion among growers about the effects of ultra violet light on resin production. Some insist it stimulates resin flow while others claim little or no effect. Two things are certain: large amounts of ultra-violet can injure the plants and you can grow high quality grass with or without ultra-violet. Another belief is that nitrogen deprivation stimulates resin production while others say that a dry medium is most important. Nitrogen uptake is regulated by the amount of moisture in the soil. Since nitrogen uptake is minimal in dry soils it really doesn't matter to the marijuana grower which is the actual mechanism. Hold watering to a minimum and keep the atmosphere as dry as possible during flowering. Cut holes in your reflectors so the humid air can escape. The dry atmosphere and soil will force more resins onto the flowering parts.

Aftern turning down the light cycle, if there is space between the plants, hang incandescents in these gaps. These will stimulate side branches to develop which will fill the available space. The output of these lights is mostly in the red part of the spectrum which will cause profuse flowering. Care should be taken that they are not hung too close to the plants, where they may cause burning of the leaves. For a 40 Watt bulb a ten inch gap will be safe— larger bulbs require more distance. For a more even light distribution, use several smaller bulbs instead of one large bulb. Heat given off by a fluorescent and incandescent bulb is about equal for equal wattages. Heat radiated by a fluorescent is spread out over the length of the tube and will not burn the leaves unless left in direct contact for several hours.

You can expect a minimum yield of about 1 ounce of pot per square foot of growing area. Large pots give fewer but taller and bushier plants. The total yield is similar for 6" to 18" pots. Eight to ten-inch pots are a good median size for high-yield, high potency grass from a moderate amount of soil. Allowing much more than 1½ feet of growing area per plant will cut down on the yield of the system.

TEMPERATURE AND HUMIDITY

Temperature control should be no problem. The plant grows well at room temperature (70 to 80 degrees during light hours, 55 to 65 degrees during darkness) and will survive in temperatures from 40 to 100 degrees.

Plant growth is closely related to temperature. The rate of photosynthesis increases until the temperature reaches about 75 to 85 degrees depending upon the variety. As the temperature rises above this level, the rate of photosynthesis slows and cannabinol resins develop. During flowering, plants grown in high temperatures (85 to 100 degrees) and low humidity will produce more resin, while during growth stages plants grow fastest at room temperatures and medium humidity. For this reason it is a good idea to start your crop so that you'll harvest during the winter months when the heat is on. Heated homes have a very dry atmosphere.

Propane catalytic heaters do a very good job of heating, are safe, clean, and increase the CO_2 content of the air. Electric and natural gas heaters also work well. Do not use kerosene or gasoline heaters. They do not burn clean and the pollutants may harm the plants.

At high temperatures and humidity, air should be allowed to circulate freely through the garden. Gardens in small, confined spaces such as closets must be opened daily or the atmosphere will become stifling and growth rate will slow down. Constant air circulation does not seem to be critical with marijuana as long as the plants obtain CO_2. If you have a large garden and there is no way for air to circulate, place a small fan in the garden. Roton makes fans for indoor gardens that are small, quiet and not affected by humidity.

CARBON DIOXIDE

Plants take in carbon dioxide (CO_2) and release oxygen (O_2) during photosynthesis, while at night plant cells respire by taking in O_2 and releasing CO_2. The net result is that much more oxygen is produced than is consumed.

Carbon dioxide concentration in the atmosphere is very low (about .03%). Around large cities it is a little higher (.04%). Plants can use much more CO than is supplied by the ordinary atmosphere. In general, the rate of photosynthesis increases in proportion to the CO_2 content of the air up to about .5% as long as there are no limiting factors such as inadequate light.

Tanks of CO_2 can be used to increase the concentration in the air. Periodically, disperse the gas above the tops of the plants. CO_2 is heavier than air and will move slowly downward.

For more information on this subject see **Lighting for Plant Growth** by Bickford and Dunn, Kent State University Press.

HYBRIDS

As you become more familiar with the marijuana plant, you may want to develop your own strain by crossing selected plants. Plant seeds from as many strains as possible. The growth patterns will vary considerably and you can select one male to go to flower and fertilize particular females. Factors for selection might be potency, yield, rate of growth or pure aesthetics. All males except the selected one must be removed before their flowers open. Place the selected females around the male plant. Periodically, shake the male or fan the air about the male's flowers. The pollen will disperse in a fine mist over the female flowers. This method should be adequate to produce enough viable seeds for your next crop. After a few generations you'll have your own strain, well suited to its environment and your taste.

Hermaphroditic plants are not unusual with marijuana. Some are genetically determined (protogenous) while others are a reaction to a hostile environment (most likely the photoperiod). An irregular or prolonged photoperiod can cause this. These plants have only female flowers at first. Male flowers appear later at the top of the stem and branches. Protogenous hermaphrodites develop male and female flowers more uniformly with female flowers directly above male flowers on the same branch.

54

Hermaphrodism can be used to develop a male-free crop. All male plants must be removed before they go to pollen. Collect the male flowers from an hermaphrodite when they are a good size but have not yet opened. Store the flowers in sealed vials. In a few days they will open. Apply the pollen with a Q-tip, artists brush, toothpick, etc. over the stigmas (white "V") on another female's flowers. Wait a few weeks until the seeds are full and have good color before harvesting. The next generation will be all females or all females and hermaphrodites (from protogenous mates).

More serious growers can try grafting hops plants to marijuana stalks to produce a possibly legal plant, using growth hormones such as gibberilic acid, or mutating to polyploids using colchicine or other chemicals. Methods for these are discussed in the following:

The Cultivator's Handbook of Marijuana by Bill Drake

Super Grass Growers Guide by Mary Jane Superweed (Stone Kingdom)

Bark Leaf - (Summer 1972) - Available from - Church of the Tree of Life, 451 Columbus Avenue, San Francisco, California 94133

Hops seeds can be purchased from:
Redwood City Seed Co.
PO Box 361
Redwood City, California

Hops cuttings from:
Wine and the People
1140 University Avenue
Berkeley, California 94707

CURING

All leaves must be thoroughly dried for comfortable smoking and full potency. The THC in fresh grass is mostly present in the form of non-psychoactive tetrahydrocannibinolic acid. Upon drying the acid is converted to THC by decarboxylization.

Single leaves can be dried by placing them in a pan on a hot radiator or in the sun. A quicker method is to preheat your oven to 150 degrees. Place a single layer of leaves on a cookie sheet, pan or aluminum foil, turn off the oven and place inside. In five to fifteen minutes the leaves will be dry and will crumble easily between your fingers. If not dry remove the grass from the oven and repeat preheating and drying. Another way is to hang plants intact, upside down above a radiator, space heater, or in the sun. Some of the resin contained in the stem will ooze onto the leaves. It will take three to ten days to dry completely, depending upon the humidity and other factors. The potency of the grass varies in different parts of the plant. Potency increases from bottom to top. The small leaves on the branches are more potent than the large leaves on the main stem and the flowering parts are the most potent of all. The female plant is almost always considerably more potent than the male. The best part is the flowering top of the female plant and the worst (which is really not bad) is the large leaves at the base of the male.

LARGE SYSTEMS

With the price of grass what it is today, some of you may want to undertake growing on a large scale. To get the highest yield for the smallest investment requires a conservation of light and soil. During the first few months of growth the plants need much less soil and garden space than they do when they are older. You can design a system that will produce large, mature plants to harvest every month by having in each system six sub-systems of plants at different growth stages. For example, fifty plants need a minimum of fifty square feet of growing room when mature, but during the first month they will fit inside of two square feet. During the second month they will need approximately six square feet.

If the plants are started in large pots, the pots themselves take up most of the room. This wastes light and soil on empty space. By rotating plants into bigger gardens and successively larger pots, you get the highest yield from a minimum investment. Transplanting to larger pots is easy. The root systems quickly fill the pots and the plants can be removed intact with all the soil adhering to the roots. This is done by turning the pot upside down, placing the base of the plant stem between index and middle finger, then tapping the bottom edge of the pot against a sturdy hard surface. The plant will pop out of the pot.

For smaller gardens use industrial type light fixtures. Larger systems should have single tubes, evenly spaced, and mounted on plywood. Big systems can get very heavy because of the weight of the transformers. It is more convenient and cheaper if you don't buy fixtures, but only the end sockets and transformers. Mount the transformers separately and run extension wires to the light system. With only the sockets and tubes mounted on plywood, the lights are easily raised and there is less weight for the walls and ceiling to support.

For larger systems it is better to use very high output tubes. These have a much higher intensity than regular fluorescent tubes, and their effective distance is so much greater that you need fewer tubes and can place them further apart. The closer the tubes are placed to each other the less efficient the lights are. Light from one tube meets light from another and is converted to heat rather than as usable light.

It is well worth it to grow all female crops either by taking cuttings or by hybridizing with hermaphrodites when building these systems.

A three garden, two month system is given as an example but the idea can be extended to a six garden, one month system.

A. The First Two Months-Plants are started in sixty-five 4 inch pots within approximately eight square feet. Using 20 Watts of light per square foot, (PSF) you are using 160 Watts from two 8 foot tubes (72-80 Watts each).

B. The Third and Fourth Months-Transplant to 6 - 8 inch pots. The system uses approximately 32 square feet. Using 20 Watts PSF you are drawing 640 Watts from eight 8 foot tubes or 3 Very High Output Tubes (215 Watts each).

C. Fifth and Sixth Months-Option to transplant to 10 - 14 inch pots within approximately seventy square feet. Using 20 Watts PSF you are drawing 1400 Watts from seventeen 8 foot tubes or 7 VHO Tubes.

See Light Charts pages 90-91.

MAINTAINANCE AND
RESTARTING

Periodically you should clean the tubes and reflectors to remove dust and grime or else the amount of visible light produced will be cut. Most fluorescents lose about thirty percent of their effective power after about a year of use. They should be replaced when dark rings appear at the tube ends. Replace incandescents after five hundred light hours.

Don't smoke around the plants. Heavy concentrations of tobacco smoke are harmful to marijuana, especially male plants.

Visiting your garden will be good for both you and your plants. You'll provide them with CO_2 and they'll provide you with oxygen rich air.

To start a new crop it is best to begin with fresh soil, especially if you had been using a system with smaller pots and frequent fertilization. A buildup of toxic salts can harm the new plants. To salvage large quantities of soil, remove the top two-inch layer of soil, which contain most of the harmful salts. Treat the rest with a trace element mixture, add fertilizer and fresh soil. Thoroughly mix and repot in clean, sterile pots.

INSECTS AND DISEASES

The indoor garden is an ideal habitat for plant pests. There should be little chance of any problems if you start with sterilized soil and keep the garden segregated from other plants. Before planting make sure that none of your houseplants are infested.

Over-watering often causes plants to lose their vitality, develop drooping and spotted leaves, and sometimes succumb to fungus or stem rot. Stem rot appears as a brown or black discoloration at the base of the stem and is soft or mushy to the touch. To correct this allow the soil to dry more before each watering and be sure to water around the stem, not on it. Wipe fungus and stem rot off the plants and treat them with a fungicide.

Spider mites and false spider mites are the most common and destructive pests. Both species are barely visible to the naked eye and usually are well established before you discover them. First indications are chlorotic or whitish leaves or bronzing of edges along the veins. Webs form at the internodes of the stem and along branches. The cyclamen mites are oval, tan to black or semitransparent. Eggs are white and laid along veins on the underside of leaves. False spider mites are bright red. You can usually see mites as tiny specks if you look up at the light system from the underside of leaves.

Mites are difficult to eliminate. If only a few plants are infested, remove and destroy them immediately. The other plants must be treated with an insecticide such as Malathion. Malathion is an organic phosphate which is effective but very toxic. However, it breaks down chemically and is metabolized into harmless chemicals after 14 days. Do not harvest in less than 14 days after spraying.

When using Malathion add one-half teaspoon of mild detergent (not soap) to each gallon of solution. The detergent will help spread the insecticide more thoroughly over the plant. If the plants are large spray the entire plant, especially the undersides of leaves and soil surfaces. The spray kills the adults but is not effective against their eggs. Repeat this application weekly for the next few weeks.

Be extremely cautious when using insecticides. You are going to smoke or ingest the plant and don't want to poison yourself along with the insects. There are a number of insecticides such as Diazinon and Malathion on the market which are safe when used as directed. The label will list precautions and give time periods for degrading before consumption. If you have a pest problem which we haven't described, your local nurseryman should be able to prescribe the proper treatment. Smaller plants should be dunked into the solution, which is the surest way to kill pests. If the plants are not heavily infested wash them in soapy water; one quarter pound pure soap (such as Ivory Flakes) to one gallon tepid water. Mix the soap thoroughly into the water. Then, cover the top of the pot with foil or newspaper, invert it, and dip it several times. Let it drip dry, and then rinse with clear water. The dunking procedure may have to be used repeatedly since is it almost impossible to wash all the mites off.

Mealy Bugs are larger (about 3/16") and white. They are usually found on the underside of leaves or near the stem internodes. The eggs are contained in a white cotton-like or waxy material at the stem internodes or leaf axils. The infested plants will need more frequent watering and have a weakened appearance.

Aphids are about 1/16" long and are green, red, pink or black. They have roundish bodies with long legs and antennae. Some species have wings. They congregate on the undersides of leaves; especially young tender leaves. Growth becomes stunted and leaves curled or distorted. Mealy bugs and aphids are not as common a problem as mites, and are easier to deal with. Remove plants infested with aphids and mealy bugs from the garden. Dunk them into a solution of 1/4 lb. soap per gallon of tepid water. Use a cloth and go over the underside of the leaves and along the stem, or use rubbing alcohol applied with a Q-tip to remove the pests. When using Malathion, one application to the whole crop is usually enough to prevent these pests from recurring.

Whiteflies are white (obviously) and about 1/16" long. The young appear as pale green or yellow scales. Usually you don't see whiteflies until the plants are moved. Then the adults take off and it looks like a small snowstorm. Plant growth is slow and leaves are often sticky with the insects' excretions. A thorough spraying with Malathion will usually get rid of whiteflies.

For further information on pest control there are two books which we recommend:

The Natural Way To Pest-Free Gardening by Jack Kramer New York City - Charles Scribner's and Sons - 1972.

Organic Way To Plant Protection, Emmaus Pa - Rodale Books, Inc. 1966.

Outdoor Cultivation

THE OUTDOOR GARDEN

Marijuana is usually an annual plant. This means that the life expectancy of the plant is based on the length of the growing season. The longer the growing season, the better the quality and larger the quantity of your crop.

Marijuana should be planted outdoors two weeks after the last threat of frost, and should be harvested before the first fall frost. You can find the approximate dates for your area by consulting experienced growers, nurserymen or the local Agricultural Service or County Agent.

Some fields are warmer than others in the same area because of the way they lie, wind, and snow cover conditions. Northern slopes are the coldest, and receive the least light. Southern slopes receive the most light and are generally the warmest. Eastern slopes are shaded in the afternoon and Western slopes are shaded in the morning. The steeper the slope the more pronounced is the shading.

PRECAUTION

Naturally you will want to be careful where you grow your crops. Make sure that there is no visible access from a road or well used path. Since marijuana may grow to twenty feet (depending upon variety, length of growing season, soil conditions and light) it might be best to intersperse it with other tall plants such as staked tomatoes, corn and sunflowers. Find out what kind of fields the growers in your area are using. An area that grows over with tall weeds will most likely grow good grass if you start the marijuana before the weeds come up.

An ideal planting area is an open clearing in a woodland not frequented by the general public. The clearing should be located so that the plants get at least eight hours a day of direct sun. Other possibilities are clearings on mountains, depressions in fields or clearings in giant fields not under aerial surveillance.

Remember that grass cannot be easily moved once it is planted and that it will probably ramain there for at least four months.

There have been a number of incidents of hunters discovering patches of marijuana and reporting it to the law. Try not to plant on land frequented by hunters.

GROWING CONDITIONS

Marijuana likes as much sun as it can get, and a moist but well drained soil. It does not do well in swampy and clay soils. The soil should be high in nitrogen and potassium and medium in phosphorous. The pH should be at least 5.5; it will do better at 6.5 - 7.5.

At least 2 months before planting you should test and adjust the soil. Needed nutrients should be added to the soil at least a month before planting for best results. This gives the fertilizer time to dissolve and become available to the plants. The pH can be raised by adding ground limestone, dolomite limestone, hydrated lime, marl, or ground sea shells.

Sandy and loamy soil can be conditioned just by adding fertilizer and making pH adjustments. Nurseries carry several different fertilizer mixes. Select the one closest to your needs as determined by the soil tests. The Agriculture Service and County Agents will do these tests for you.

Turn and loosen the soil and break up large clods of earth. Clear all ground cover near the spot where you are planting. Add fertilizer and work it into the ground. If it rains frequently in your area, the fertilizer will soak into the ground by itself. If not, water the area so that it dissolves.

Clay soils can be adjusted by working in straw, manure, leaves and stalks, compost, used kitty litter, or construction sand. These help to keep the soil loose and aerated.

Swampy areas can be adjusted by building planting mounds about one foot high and one foot across. The mounds will have better drainage than the surrounding soil and they will not become waterlogged.

If the soil is very bad and you are only growing a small patch, there are several other ways of changing soil conditions:

1. Buy topsoil, and place it in holes where you are going to plant. This is only good for small gardens since it is laborious and expensive.

2. Dig a hole one foot deep and one foot wide. Fill it six inches deep with manure or compost sprinkled with lime. Fill the remainder of the hole with soil.

3. Use a self contained planting pot as described in Transplanting.

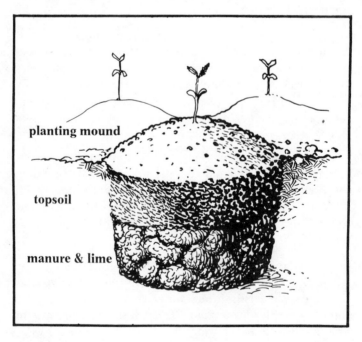

SEEDLINGS AND TRANSPLANTING

To get a longer season you can start seeds indoors and transplant them outside after the threat of frost has passed. This is especially helpful in the Northern U.S. where the growing season is short. Seeds can be started as much as two months before the season begins. There are several methods for starting seeds:

1. Planting Pellets. These are one and a half inch pellets which expand when they come in contact with water. They come in several different pH levels. Get either a 6.5 or 7. These are the easiest units for starting seedling. Just follow the directions on the package. They should be used only if you are planning to plant within a month.

2. Planting Pots. These pots are made of compressed peat moss. They come in all sizes, but the best one is probably 2" x 2". Fill with one of the soil mixtures described in Indoor Cultivation. Try to prepare from the same soil to which the plants will be removed later. Plant several seeds in each pot and thin to one plant per pot. When you are ready to transplant outdoors just dig a hole in the ground and put the planting pot in it. The roots will grow through the pot and it will eventually disintegrate. Tin cans can be used in place of planting pots. Make sure that the cans have drainage holes and the sides are scored so that the roots can grow out ot them. Do no use aluminum cans. They won't disintegrate and the plants' roots will be trapped.

3. Seed Trays. Seed trays are the most economical way of starting large numbers of seedlings, but the plants' roots may be damaged when you transplant. Fill plastic planting trays with one of the mixtures described in Indoor Cultivation. Sow one seed every inch, but thin to one plant every two inches when they begin to interfere with each other. When you are ready to transplant them slice the soil into squares and plant outdoors. To prevent shock treat the plants with Transplantone or Rootone according to package directions.

4. Self Contained Soil Unit. This method should be used only when the soil is unsuitable for adjustment. Use institution-size cans, large juice cans or bushel baskets, filled with three inches of vermiculite or perlite mixed well with a slow release fertilizer, and then fill it the rest of the way with a mixture of soil with perlite, vermiculite or sand. A mixture of soil and manure or humus with potassium added can also be used. Holes should be punched in the bottom of the can for drainage. When you are ready to plant outdoors, put the can in a hole in the ground.

Use the same methods in cultivating these plants indoors as you would if these plants were to remain indoors permanently. If you are planning to keep the plants indoors for more than a month they have to be introduced to the sun's intensity gradually. The plants need at least 40 Watts of fluorescent light per square foot to avoid shock. This will also build up the sugar supply to help the plant withstand transplant shock. Other ways of avoiding shock are by putting trays containing the seedlings outdoors in a partially sunny area for a few days prior to planting, putting a light shading cloth over the plants for a few days after they are transplanted, or by keeping the plants at a very bright window for a week before transplanting. Using full spectrum light (Vita-Light, Optima and Naturescent) will also minimize shock.

If you have indoor plants already growing you can clip shoots about 3 inches from the growing tip and put each of them in one of the containers mentioned previously. They will quickly develop roots and start growing into new plants. This is a good method of obtaining high quality transplant stock.

The night before you transplant, water both the plant and the soil to which you are going to transplant. Also, to prevent shock, transplants should be made to and from soils with the same chemical and textural characteristics.

Plant on a cloudy or drizzly day or late in the afternoon. Never plant or transplant on a bright, sunny day. The sun's energy is too much for the plants to take at first.

SPACING

Marijuana is very adaptable and can be grown as close together as 15 inches between rows with plants every 6 inches. Plants grown this way will not be as bushy as plants grown further apart. Plants grown six feet apart will be very tall and bushy because they get plenty of sun and have less competition for nutrients. Spacing rows about 24 inches apart with plants about 15 inches apart seems to be the most efficient method of utilizing the area. Plants will be bushy, tall and easy to harvest.

In order to catch as much sun as possible rows should run north to south—perpendicular to the course of the sun.

WATER

Marijuana cannot grow (or live) in an environment in which it cannot find water. It sends down a tap root which may grow half as long as the plant. Often marijuana can be found near the banks of streams in dryer areas. Cultivated fields supply enough water naturally or through irrigation. Some growers in remote areas use portable water pumps run by engines or generators. Digging a hole in which the generator can be run and stored will muffle sound and keep the machinery in better condition. Make sure not to overwater your plants. Keep the ground moist, but not waterlogged.

CARE

Grass is at its most vulnerable stage right after germination. The seedlings have a tendency to fall over in rain and wind. Usually they can overcome their crises. If you have started seedlings indoors you will be past the critical stage by the time you transplant.

One-and-a-half to two months after germination you will have to decide whether to clip the tops to make the plant bush, or to let it grow straight up and bush on its own. Letting the grass grow straight will allow it to produce more weed, but bushy plants are harder to detect. If you want the plants to bush, cut the main stem about 3 inches from the top when the plant is between two and three feet tall. Very long secondary branches should also be trimmed. The clipped tops can be dried and smoked, of course, or they can be rooted. This process should be repeated if the plant starts growing tall again.

If you have prepared the soil properly you will not need to fertilize during the growing season. It is a good idea to check the plants periodically. If the plants seem to have any deficencies, add the proper nutrients. If the plants are not growing quickly, make sure they do not have too much competion for sunlight. If the plants are too close together they can be trimmed or pulled. If crowding is not the problem pH probably is. Test the pH and make the proper adjustments.

FLOWERING AND HARVESTING

The plants will begin to flower in late August or early September. When total daylight hours fall below 13 - 15 hours a day (depending on variety) the plants are triggered into the reproductive cycle.

If you have a long growing season and secure conditions, pick the flower buds off. The plant will send up new buds. As long as the plant continues to send them up you can clip them off. Some say that this increases the potency. It surely increases the yield.

Many farmers throughout the world bend the stem of each plant sharply at a point way down. The plants are left this way for several days after which the sun dried tops are harvested. The bend cuts off circulation between the upper and lower parts of the plant. Cannabinol resins cannot flow back down past the bend. Furthermore the shock of bending apparently drives the resins in the portion of the stem just above the bend towards the flower tops.

Another technique used is to bend the tops more or less horizontally so that they snap, but do not crease. The tops draw some liquids from the base of the plant, but not enough to stop them from wilting within ten days. People who use this method claim it increases potency significantly.

In many places, most notably in India and Pakistan, farmers make the practice of destroying all male plants as soon as their gender becomes determinable. This is done to prevent their maturation and the pollination of the females. It has been found that a loss of cannabinol resin often occurs in the female flowertops shortly after pollination.

If your growing season is short let the plants flower and harvest them before frost. Some claim that marijuana is at its potency peak at this time. Others claim that marijuana is at its most potent state about two to ten days after it starts to flower.

If you wait till the seeds mature and drop off the plant, you may have a crop next year without planting. It is almost impossible to get rid of marijuana once it have become indigenous to the area. The Federal Government has gone so far as to suggest that farmers in Iowa and Kansas napalm or herbicide their fields.

Marijuana can be harvested by pulling up the whole plant including the roots, by chopping it off about ½ way up the stem, or by picking each part of the plant separately.

Depending on cultivation methods and environmental conditions you should harvest between 1000-5000 pounds per acre (43,000 square feet).

PLANT PESTS — OUTDOORS

Several different kinds of insects like to eat or suck on marijuana. Several methods can be used to prevent them from getting at it. Companion Planting of onions, garlic, chives, savory, thyme or marigolds keeps some insects away. Inter-crop one of these with your marijuana.

Predatory insects such as praying mantis', lady bugs, and lacewings eat insects which attact marijuana. They can be purchased from commercial hatcheries. Do not spray plants with insecticides when predators are present.

Botanical repellants, naturally occurring insecticides which have been concentrated, can be used in spray form. They are not persistant, that is, they do not build up in living tissue, but they are poisons. Pyrethrums and Rotenone are the ones most often used.

Your plants are more likely to be attacked by foraging animals. Blood meal placed on the ground near the garden will keep deer away. Chimes, bells and scarecrows will keep foraging animals away somewhat. Fences can also be used successfully to keep hungry animals from your garden.

For further information on pest control there are two books which we recommend:

The Natural Way to Pest-Free Gardening by Jack Kramer, New York City, Charles Scribner's Sons, 1972.

Organic Way to Plant Protection, Emmaus Pa., Rodale Books, Inc., 1966.

EXPERIMENTS

Recently there have been claims made regarding the use of special techniques and tricks which some say increase the potency of marijuana. We have not investigated most of these but feel that we should pass them on for others to experiment with.

1) Blue Mold—After harvesting, while the grass is still wet, place it in a plastic bag together with blue mold frequently found on citrus fruits—especially oranges. When the mold has covered the grass dry the grass out. The technique is supposed to increase potency.

2) Burying—Place wet grass in a plastic bag and bury it in the ground for 3 weeks to 6 months, or at least until it is moldy. Both the mold's action on the grass and curing in an enclosed atmosphere are reasons given for the supposedly increased potency.

3) Carbon Monoxide (CO)—Higher than normal levels of CO supposedly cause sex reversal in some cases. CO is a toxic, odorless, colorless gas produced by incomplete oxidation. It is one of the pollutants found in car exhausts and its action on plants was first noticed along highways.

4) Carbon Dioxide (CO_2)—There have been claims that wet grass placed in an Oxygen-free atmosphere for 75 minutes at about 100° (212°F) will undergo chemical changes which increase potency. Atmosphere in a container is made oxygen free by replacing the air with CO_2, Nitrogen, or Nitrous Oxide.

5) Dry Ice (CO_2)—Place wet grass in a plastic bag or coffee can with dry ice. Put the bag in the refrigerator or freezer so that the ice takes longer to evaporate. Make a few small holes in the container for CO_2 to escape. This is said to increase potency.

6) High Altitude—Some claim that the best grass is grown at high altitudes. Reasons given are conducive air, soil, light, temperature, lower humidity, or a combination of these things.

7) Iron Nails—Nails placed in the soil with their points down are supposed to pick up static electricity and help the plants to grow by increasing the charge of the soil.

8) Light Interruption-A number of experiments have been done which indicate that the percentage of females can be increased by breaking the night (dark) cycle with 2 hours of light in the middle or 30 minutes of light every two hours.

9) Plants that have flowered or are losing vitality can be forced back into an early growth stage by leaving lights on continuously for 24-72 hours.

10) Music—There have been experiments on plant growth which indicate that plants respond to music. Some music increases plant growth, although several growers have stated that rock can kill them. A recording of music and sound supposedly conducive to plant growth is available from Edmund Scientific Corp., 600 Edscorp. Bldg. Barrington, N.J. 08007.

11) Negative ions—Experiments have shown up to a 50% increase in plant growth in areas containing heavy negative ion concentrations. More information is available in the September 1973 Smithsonian Magazine.

12) Mutations and Polyploidism—Mutations and polyploidism occur when seeds or plants are exposed to radiation, colchicine, mustard gas or chloral hydrate. Polyploidism is the doubling or tripling of the genetic material in a cell by breaking the chromosomes and will increase the vigor and/or potency of marijuana.

13) Thumbtacks—Placing thumbtacks into the stem at the base of the plant has been said to increase the resin production of the plant. This may have something to do with the plant's ability to deliver water to its upper parts.

14) Ultra-Violet—Some bands of UV light may increase potency.

15) Psychic Vibrations—Backster's polygraph experiments show that plants respond to emotional vibrations around them. Others claims that this determines potency and growth rate to some extent.

16) Lunar Phase—There are two schools of thought regarding the appropriate phase of the moon under which

to plant marijuana. Some farmers say that the best time is during the full moon. Others believe that it is better to plant during the new moon, and harvest during the full moon. Both schools agree that the moon should be in a water sign (Cancer, Pisces, or Scorpio) at planting time.

17) Marijuana plants are said to do poorly if grass is smoked or dried near the growing plant.

18) Mascots—Some say plants grow better when there are small animals living in the garden. Camellions, toads, frogs and lizards are among the creatures recommended. Be sure to supply the proper food for the animals.

If you try any of these processes please let us know how they work so we can pass the information on to others. Address all correspondence to Ed Rosenthal, c/o Clearlight Company, P.O. Box 1887, San Francisco, California 94101.

Clearlight Growing Systems

Clearlight Co. has everything you need to grow your own garden. Complete units, components, and accessories. We stock VHO fixtures and most kinds of fluorescent growing tubes, testing kits, soil-less mixtures, fertilizers, automatic watering units and timers. All our items are competitively priced.

We also offer a shopper's service—If we don't stock an item we will tell you where to get it, or get it for you (no extra service charge).

SEND FOR YOUR CLEARLIGHT CATALOG TODAY.

Here's what others say about us:

"Ed Rosenthal wants cannabis fanciers to stop seeing their dealers. Not that he's a narc or anything—to the contrary. He just has a proposition that he thinks will help de-commercialize the grass market.

"Rosenthal is offering the first grow-it-yourself grass environment. His six foot high constructions are supposed to be producing enough marijuana to keep a stash perpetually self-sufficient after about four months."

Rolling Stone 6/22/72

"Since Ed had been raving about the extraordinary quality of his home grown product, I called together an impartial and fairly knowledgeable panel to test his claim. Participants reacted with extreme enthusiasm, a majority insisting it was the best grass they had ever smoked. One in particular waxed ecstatic about the marvelous and amazing hallucinogenic qualities of the huge-leafed organic grass, but I can't vouch for her; she may have been stoned at the time."

New York *Village Voice* 6/24/71

"At last, somebody has come up with a system for growing marijuana at home which produces a superior quality smoking dope for all aficionados . . . In fact, any of the components involved in Ed's greenhouse system—soil, fertilizer, etc.—can be had from him at less cost than your local retail outfit if you want to build it yourself. You see, he's something of a missionary on the subject of psychedelics . . ."

Screw Magazine 8/30/71

Please send me my free Clearlight Catalog today.

NAME_____

STREET_____

CITY_____ STATE_____ ZIP_____

Send to Clearlight Company, P.O. Box 1887, San Francisco, Ca.
94101

Model CGS A 50 inches long, 6 feet high, 13 inches wide.
Comes with 14 pots, and 2 fluorescent
tubes (Vita-Lite). After 4 months, this
system will supply enough grass for two
moderate smokers and their growing circle
of friends.

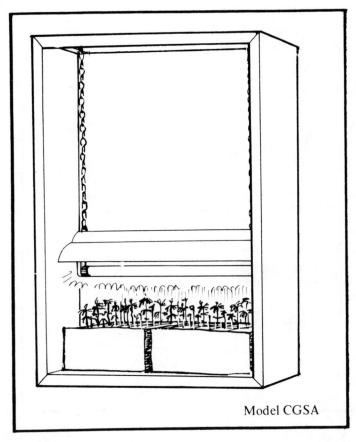

Model CGSA

Model CGS B 50 inches long, 6 feet high, 27 inches wide. Comes with 28 pots and four 4 foot fluorescent bulbs. This system will grow enough grass for 2 heavy users and their friends. **CLEARLIGHT'S MOST POPULAR MODEL**

Model CGS C 50 inches long, 49 inches wide, 6 feet high. Comes with 56 pots and 8 four foot fluorescent bulbs. This model produces enough pot for a small commune or some very heavy dopers.

Model CGSC

Light
Charts

LIGHT CHARTS

Incandescent Light Distribution
[Intensity]

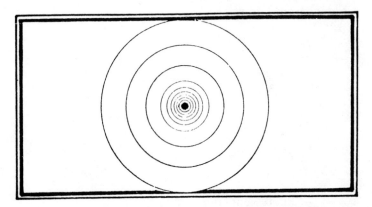

Fluorescent Light Distribution
[Intensity]

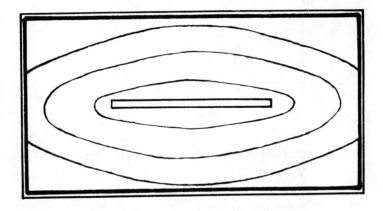

**Light Output, In Foot-Camdles [fc],
From Two Standard 40-Watt
Fluor Escent Lamps**

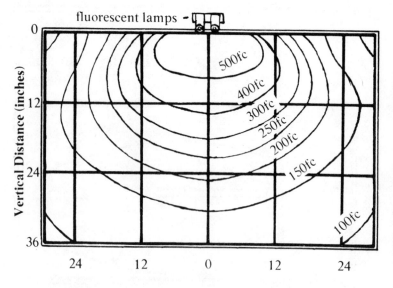

Horizontal Distance (inches)

90

Light Output, In Foot-Candles [fc], From Two VHO 8' Fluorescent Lamps

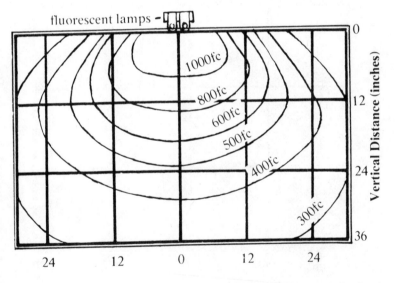

Horizontal Distance (inches)

**Comparison Of
Wide Spectrum [---]
and Standard Gro-Lux**

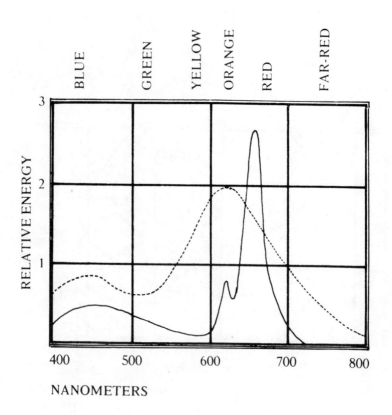

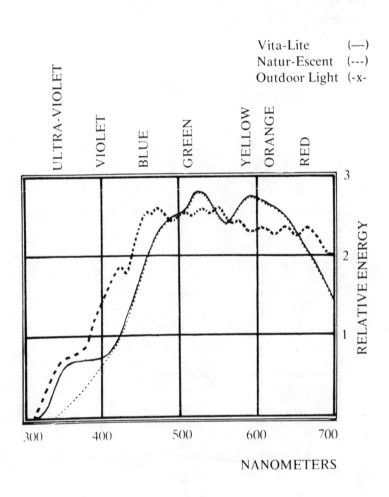

**Fluorescent Tube
Relative Energy
Output**

Vita-Lite   (—)
Natur-Escent (---)
Outdoor Light (-x-

ULTRA-VIOLET

VIOLET

BLUE

GREEN

YELLOW

ORANGE

RED

RELATIVE ENERGY

3

2

1

300   400   500   600   700

NANOMETERS

Chlorosynthesis   And
Photosynthesis Curves

Chlorosynthesis Curve   (---)
Photosynthesis Curve   (—)

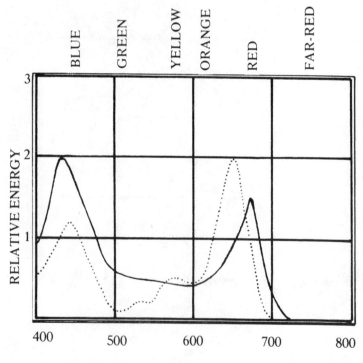

ERRATA

Page	Chap.	Line	Change	To
21	1	9	(H2O)	$(H_3O)^+$
21	1	9	(OH_2)	$(OH)^-$
22	3	7	mix to acidic	mix too acidic
37	3	2	maintaining	determining
			Add.. after single compon-ent fert..	Test results will not be accurate after fertilizing; the test results being much too high.
38	3		Delete sentence.. Use one tablesp..	(Already said on 37)
48	2	4	photosynthesis	chlorosynthesis
73	1	1	August	September
73	1	2	September	October